Sensors and GPS for Drones and Quadcopters

Impressum:

Bibliographische Information der Deutschen Nationalbibliothek:
Die Deutsche Nationalbibliothek verzeichnet diese Publikation in der
Deutschen Nationalbibliografie; detaillierte bibliografische Daten
sind im Internet über http://dnb.dnb.de abrufbar.

© 2022 Roland Büchi
Herstellung und Verlag: BoD - Books on Demand, Norderstedt

ISBN: 978-3-7557-5578-4

Content

1. Introduction and functionality

1.1 Sensors for quadcopters and drones

The name of this model construction division, which was completely new just a few years ago, has developed. At the beginning there was the term 'quadcopter'. Since many model makers also build systems with six or eight propellers, there are also the terms 'hexacopter' and 'octocopter'. These different designs are also referred to with the general term of the multicopter.

The term 'drone' has also been used for these systems since around 2010. This term is originally known for military flight systems. This refers to unmanned flying objects that can fly autonomously via GPS or also remotely. It is important that they are unmanned, i.e. that no pilot is sitting in the cockpit. In principle, this can mean the systems of all possible flight principles, i.e. in addition to quad- and multicopters also fixed-wing aircraft, as well as helicopters. In colloquial terms, however, the term drone has now almost become a synonym for quadcopter or general multicopter. This book only deals with this colloquial term drone, i.e. with the quad- or multicopters. The unmanned fixed-wing aircraft are therefore not dealt with here.

The sensors used have also evolved. The first quadcopters only had gyros in two or three axes. This resulted in high demands on the pilot of the drone. The acceleration sensors were added later, the combination allowed an automatic angular control. Then came the compass, the air pressure sensor and the GPS. Today all types and other sensors are used, which greatly simplifies the control of the drones. The sensors are the key for a simple control. The aim of this booklet is to explain their function in context with drones.

1.2 Steering mechanism and technical background

Quadrocopters are aircraft with four propellers. They have the same control capabilities as helicopters. Figure 1 illustrates this. The stick assignment of the remote control, as shown in Figure 2, is most commonly selected. However there are also model pilots who swap the left and right sides.

Figure 1: Steering possibilities.

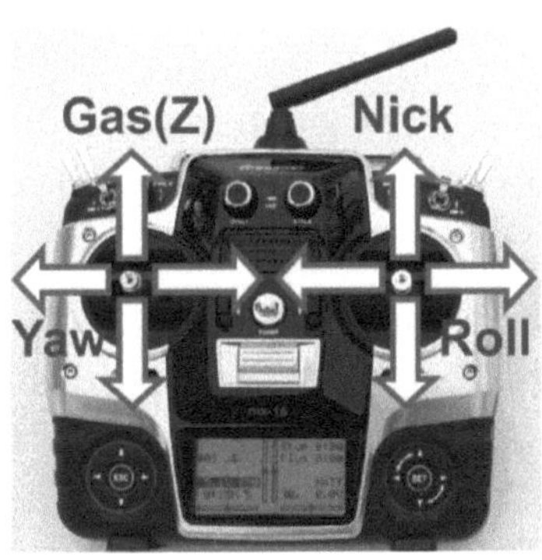

Figure 2: Stick assignment.

'Nick' describes the tilting forward and backward. For that purpose, the stick of the remote control needs to be moved upwards (tilting forward) and downwards (tilting backward).
'Roll' describes the tilting to the left and right. The stick needs to be moved to the left and the right side.
'Yaw' describes the rotation around the vertical axis (z). The left stick needs to be moved to the left (counterclockwise yaw, view from top side) or the right (clockwise yaw, view from top side).
'Gas' describes the movement along the vertical axis (z). If the left stick is moved down, it means descent flight, and if the left stick is moved up into the full throttle position, it means climb flight.

1.3 Physical movement

The immediate question is now how a quadrocopter can be controlled physically with the above functions. A helicopter will again serve as a comparison.

'Nick' and 'Roll' are there realized with a so-called swash plate. This provides at the end an angle-shift of the main rotor force axis to the fuselage. 'Gas' is provided by 'pitching', which is achieved by changing the pitch of the rotor blades. 'Yaw' is realized by a change in speed of the tail rotor. Some models also reach yaw by pitching the tail rotor blades.

Anyone who has ever built and flown helicopters knows that this requires quite a complex mechanism. A hard landing is rarely forgiven: bent rods, ragged ball heads and expensive repairs are the consequence. Many have thus abandoned the model helicopter hobby, the so-called pinnacle of model aircraft.

Quadrocopters, which – as mentioned above – have the same movement possibilities as helicopters, in contrast stand out by virtue of their much simpler and thereby massively less sensitive mechanics: There are four motors, which are rigidly connected with two right- and two left-rotating propellers – and that's all. Everything else is provided by a little electronic control board or a flight controller. Figure 3 illustrates this.

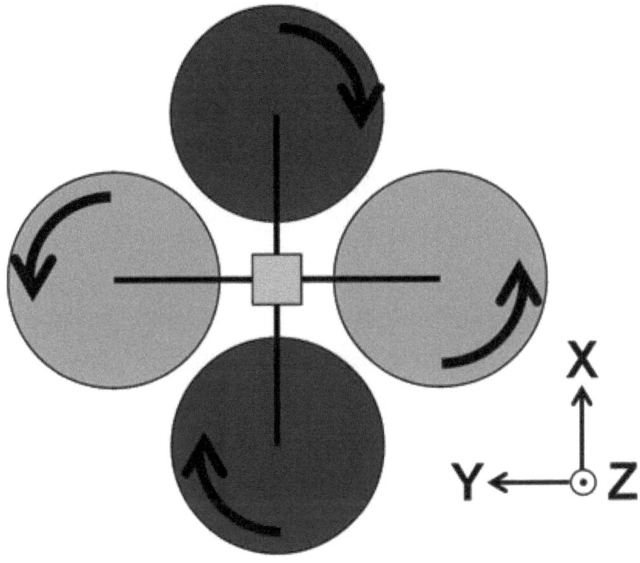

Figure 3: Two left- and right- rotating propellers, view from top side.

'Nick' is physically achieved by a change in speed of the upper and lower propeller (see figure 3). To move the quadcopter in the X direction, the lower propeller is turning faster and the upper one slower. Thus, an inclination in the direction of the x-axis is achieved.

'Roll' is achieved by a change in speed of the left and right propeller. A movement in the Y direction requires a higher speed of the right and a lower speed of the left propeller.

The main rotor of a helicopter produces a torque about the vertical axis (z) because of its twisting. The tail rotor serves to compensate for that torque. The two right- and left-rotating propellers of the quadrocopter do this job instead. Thus a tail rotor is not needed. 'Yaw' is achieved by ensuring that both left and right propellers have a different speed than both upper and lower ones. A counter-clockwise yaw (viewed from above) requires a higher speed of the upper and lower propeller and a lower speed of the left and right one.

A change in 'Gas' requires a change in the speed of all propellers together. During the climb flight, all propellers have a higher speed.

As mentioned above, quadrocopters and helicopters are controlled by the same functions and also have the same possibilities of movement – almost. Because of the control over the speed of the propellers it is not possible to fly stably overhead and to 'mow the lawn', as some pilots demonstrate with their pitch-controlled helicopters.

Loopings, on the other hand, are also possible with quadcopters, they are flown in the same way as with a wing plane, whereby the gas is slightly removed shortly before the apex and then strongly tightened again for the subsequent stabilization in the hover position. However, this presupposes that the operation allows this, because the sensor support at many systems ensures that the maximum angular position is limited by the software. However, many ready-to-fly quadcopters support automatic loops, which can be triggered by pressing a switch or a button.

1.4   Flight in ‚x'‑ or ‚+'‑ configuration

Since a quadrocopter is constructed so perfectly symmetrically, the question of where the front is, is justified. For most systems it is as shown in figure 4. This flight configuration is called the 'x' configuration. So here is not one boom at the front, but the middle between two booms. This configuration is also the standard for a photo flight, because it allows the camera to be mounted facing forwards without the annoying boom and motor / propeller combination. It is then mounted with or without an existing gimbal, a camera holder that can rotate and swivel. In the normal position, it is inclined slightly downwards so that the front propellers do not come into the picture on the left and right as far as possible. Compared to the '+' configuration shown in figure 1, in which a propeller is in front, the control board or the flight controller must carry out a so-called coordinate transformation. This also has the advantage that the model can be provided with a body, for example made of plastic or depron. On the one hand, this means that the orientation is always clear and it is also possible to fly a little further away; on the other hand, this brings the quadcopters a little closer to the 'rigth' flight models by taking creativity into account when building the fuselage. Figure 4 shows an example of a model in 'x' configuration (a model from the future?).

Figure 4: Example of a quadcopter in 'x' - configuration.

Nevertheless, in most cases, the symmetry prevents you from flying much further than 50 m without the aid of sensors and GPS. In contrast to fixed-wing aircraft, even with the orientation aid of a fuselage, it is rather difficult to recognize the orientation well at larger distances.

2. Basic sensors

The quadcopters, whether they are self-made or bought as ready-to-fly (RTF), have all installed similar components. Most flight controllers or control boards are at least equipped with the sensors necessary for pure hovering, i.e. with gyro and acceleration sensors. In addition, many have built-in compasses and air pressure sensors and also GPS sensors. With software that is becoming more and more intelligent, it is now possible to integrate many other sensors. This is shown by a large number of RTF systems in particular. If these are used correctly, it is a very big plus point in terms of user-friendliness. For example, automatic take-offs and landings can be carried out with sensors that measure the distance to the ground. This helps a lot, because as in manned flying, these are the most delicate maneuvers in quadcopter flight. Often the camera, which is usually built in anyway, with additional sensors is also used for obstacle detection. These sensors are not absolutely necessary for a quadcopter to fly. However, it goes without saying that they support the model pilot very well so that the system does not get into dangerous situations and will be damaged in the process. In particular, the many possible additional sensors are not supported by all flight controllers or RTF systems. Before deciding from which manufacturer the flight controller or the components should be obtained, you should already think about which sensors you would like to use and find out about the respective availability, for example via the Internet. Anyone who only wants to fly and take photos with the quadcopter should purchase one of the many complete systems. These usually have excellent flight characteristics. In addition, the settings and control parameters are then optimally matched to the system. Often, however, the complete systems can't be expanded with additional components. So if you want to do more than just fly and use the camera that is built in in most cases, for example if you want to install additional sensors or implement other software structures, you should rather opt for individual components and then assemble them yourself.

Figure 5 shows the components of a quadcopter that are at least required for flight. These components are also built into the RTF quadcopters, but they are a little more difficult to photograph and distinguish there. They are described in more detail in the following subsections.

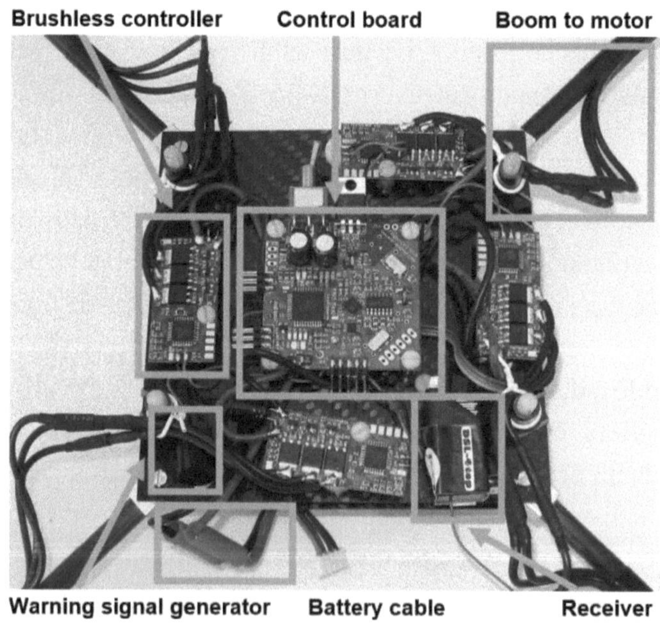

Figure 5: Basic components of a quadcopter.

2.1 Control electronics or flight controller

Remote controlled model aircraft or helicopters usually have no control board. The receiver takes over this function. It drives the servos, the gyro for the stabilization of the tail rotor and the brushless controller for the motors. In the case of a quadrocopter, the functions of the receiver alone are not sufficient to achieve all requirements. This has several reasons.

- The main reason is physics. Without additional sensor support a human being would not be able to stabilize and fly a quadrocopter. Looking at an axis (such as the pitch axis) from the side, this looks like a swing with two drives at the ends. If both motors do not produce exactly the same thrust (which they never do in practice), the axis will start to turn. To compensate for this, one would have to observe this rotation and manually reduce the gas in one of the motors.
We can illustrate a similar problem by trying to stabilize a broom which is on the palm of a hand. Balancing it with the hand works pretty well. Now we try to stabilize a 40cm long aluminum rod in the same way. This is much more difficult because the rod is smaller and quicker in its movements.
It's the same with the quadrocopter, where the spacing of the drives is usually of a similar size. It is therefore necessary to measure at least the roll and nick angles. In addition, in the yaw axis, the angular velocity (yaw rate) is measured. The control board then balances the axis through the sensor signals by itself. That means it

controls the speed of the motors, but much faster than humans could.

- As mentioned above, several motors and propellers together must execute a certain action following a stick movement. It is not the case that a stick is responsible for the movement of only one motor alone. A mixing function is therefore necessary. This is flexible enough only with a microprocessor on the control circuit board and corresponding software.

- The stabilization due to the sensors – the above-mentioned balancing – is realized by PD controllers. The P and D parameters must be adapted according to the choice of other combinations of motors and propellers or larger or smaller riggers. They play the decisive role as far as the properties of the quadrocopter – agile and fast or slow and good-natured – are concerned. Therefore, it is necessary that they can be reconfigured using the PC interface.

- A configuration with a PC interface allows also a range of other parameters such as the effect of the control stick, an emergency gas function upon a failure of the receiver signal, a battery warning of low voltage, etc.

Therefore, the control electronics board or the flight controller plays a central role in building the quadrocopters. In summary, it provides the following basic functions:

- Power of the four brushless controllers via a central switch

- Calculation of the angle in the nick and roll axis and the angular velocity in the yaw axis using the gyro and acceleration sensors
- Control of nick and roll angle and the angular velocity in the yaw axis with PD controllers (balancing)
- Calculation of motor voltages due to the control and the throttle position
- Signals for the brushless controllers
- Special functions, e.g. actions after the loss of receiver signal, battery warnings
- Read the GPS and other Sensors discussed below.

Some of the functions are absolutely necessary for a quadcopter to be able to hover and fly at all. But other functions are also important for correct functioning, for example the connection of GPS, compass sensors, air pressure sensors, or sensors for detecting the distance to the ground or obstacles. These are also covered below.

2.2 Gyro- and acceleration sensor, IMU

These are the two most important sensors, and in most quadcopters, they are integrated on the control electronics board or the flight controller, as is the case with the example in figure 5. Based on the problem of measuring the angles of the nick and roll axes and the angular speed in the yaw axis, the functionality of gyros and acceleration sensors is explained first. Then it is shown how the angles or the angular velocity can be calculated from this.

Gyro

Most of the gyros that are used with drones are implemented with MEMS 'Micro Electro Mechanical System' technology. Sometimes they are also called SMM for 'Silicon Micro Machine'. This means that the following explanations are technically all implemented on a silicon chip. Figure 6 shows a schematic diagram. Schematically, a mass is suspended there with a spring. It is called the seismic mass. This is now excited to vibrate. In the image shown here, it swings vertically.

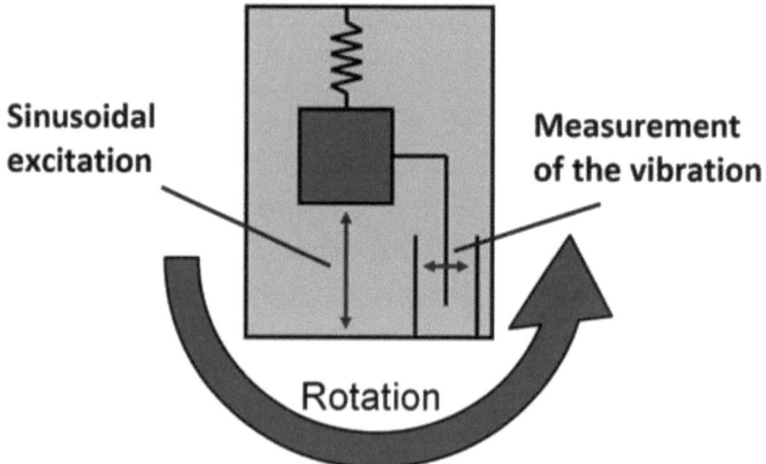

Figure 6: Functionality of a gyro, principal sketch.

The so-called gyroscopic effect now ensures that a rotation makes this mass oscillate in a horizontal direction. The oscillation is greater, the faster the rotation, i.e. the greater the angular velocity. The intensity of the oscillation and thus the angular velocity can be measured directly. This can be done capacitively and then works in such a way that the oscillating mass moves a metal plate between two capacitor plates and thus changes its capacitance. If the rotation is in the opposite direction, then the oscillation is out of phase. The direction of rotation can also be recognized in this way.

MEMS technology is now the first choice for the construction of gyro sensors. Today, however, there are still sensors available that work on the basis of piezo elements. The properties of the piezo elements are then used for both the excitation of vibrations and the measurement. However, these show a temperature drift, which is why quadcopters equipped with them often have to perform an automatic calibration routine before starting.

Acceleration sensor

Figure 7 shows an acceleration sensor. A mass (known again as seismic mass) is suspended by a damped spring in all three spatial directions. If the system is inclined, as illustrated, the acceleration of gravity g causes a shift of this mass.

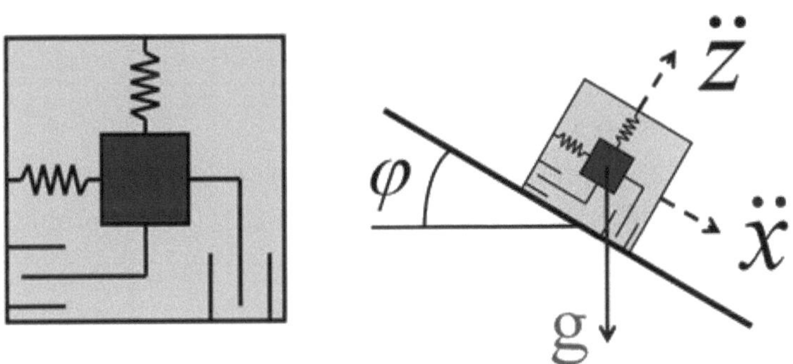

Figure 7: Acceleration sensor.

The projected components of g are now measured in the x- and z-direction. The two dots on x and z represent acceleration.

The measurement is done capacitively and works so that the moving mass moves a metal plate between two capacitor plates, thereby changing the capacities. Using some trigonometry we see:

$$\ddot{x} = g \cdot \sin(\varphi)$$

$$\ddot{z} = -g \cdot \cos(\varphi)$$

Since 'sin/cos = tan' it follows that

$$\tan(\varphi) = -\frac{\ddot{x}}{\ddot{z}}$$

For a small angle, the tangent is always the same size as the angle itself (to test: set the calculator to RAD) and so the equation simplifies to:

$$\varphi = -\frac{\ddot{x}}{\ddot{z}}$$

Acceleration sensors typically measure three dimensions on a chip, i.e. in the x-, y- and z- axis. The tilt angle in the nick axis, as described above, is measured with the x and z components, and the one in the roll axis with the y and z components.

Combination of sensors
So one could simply say: One needs an acceleration sensor for the nick and roll angle and a gyro for the angular velocity in the yaw axis.
This is correct for the yaw axis. For the nick and roll angle, however, things are a little more complicated. Using both the gyro and acceleration sensor, the calculation of the angles is possible using either the gyro or the acceleration sensor. Both, however, have disadvantages, so that only a combination of

both (called a sensor or data fusion) gives a reasonably reliable result.

A little physics may help: The angular velocity is the time derivative of the angle, in other words explains the angle change per time unit. Summing the signal of the gyro numerically, we therefore form the integral and we get the angle. This is done such that the signals are added at specified times. The problem with numerical integrations is, however, that errors add up and thus after a short time the calculated angle no longer corresponds to the measured angle.

The acceleration sensor does provide the angle, but only in so-called stationary cases. If the quadrocopter is inclined, an acceleration in the direction of the slope will occur, which is of course also measured. So the angle calculation is invalid. In addition, wind influences distort the result. Only without wind and with a movement with constant velocity (i.e. no acceleration) the angle is measured accurately. Then the above formulas are valid.

A lot of development work in the field of sensor fusion is being undertaken in order to accurately measure the angle with a combination of gyro and acceleration sensors. Model-based approaches or Kalman filters are the technical terms for this. In practice, however, a simple solution is usually implemented. The angle is integrated numerically based on the gyro sensor as described above. The signal of the acceleration sensor, which often – although not always – provides a reasonably good signal for the angle, is used as reference to match the integrated gyro signal. It thus ensures that the angle does not drift away as a result of the numerical integration.

Anyone who drives speed flights with their quadcopter may have already noticed that some systems no longer straighten up exactly after fast maneuvers or turns. The reason for this is that the acceleration sensor cannot measure the angle exactly in this flight position because of the above statements and so incorrectly adjusts the integrated gyro signal. In summary, there must therefore be a gyro sensor in every rotation direction (roll, nick and yaw), and for the adjustment of the angle of roll and nick an acceleration sensor in x, y and z in addition.

IMU, Inertial Measurement Unit

In some systems, the software for combining the gyro and acceleration sensors is implemented using the control board. Today, however, the market for products that must have exact angles is very large. These sensors are also built into smartphones or game consoles and much more, for example. In addition, the calculation of this combination is not entirely trivial, as the above example shows with the incorrect angle after fast maneuvers or curves. This is why some manufacturers have specialized in the calculation and measurement of these sensor combinations and offer complete solutions, either on circuit boards or directly on the chip. These are then called IMU, for 'Inertial Measurement Unit'. All the necessary sensors and calculations are then located on it, and the solid angles are returned as the result. Often the air pressure sensors and compasses discussed below are also integrated, perhaps even GPS. Technology is advancing here too and the solutions are, in the end, increasingly cheaper, highly integrated chips that are soldered directly onto the control circuit boards and provide very precise spatial angles or other position information at all times.

2.3   Additional sensor support

With the basic components, the quadrocopter pilot already holds in his hand a system with which he can perform, as described above, the same control features as with a helicopter. He can on the one hand operate speed flight by attacking the model at an angle, thus generating a driving direction. On the other hand, he can also levitate the model right on the spot. Well-trimmed, with the angle control of the roll and nick axes and the angular velocity control in the yaw axis, in calm conditions it is perfectly possible to let go of the remote control and watch the model slowly drifting away. And yet there are other ways to control it even better, with additional sensors. To illustrate this, figure 8, similar to figure 1 is shown again, with a quadcopter in 'x' configuration.

Figure 8: The six degrees of freedom of a drone.

Generally a body in space has six degrees of freedom. In this way, the possibilities of movement are defined. As shown in the figure, there are the three degrees of freedom of angles (nick, roll and yaw) and the three degrees of freedom of axes (x, y and z).

Using only the basic components, relatively few degrees of freedom are actually controlled by the model itself. These are only the two angular degrees of freedom nick and roll. The fun of flying is therefore due to the fact that the quadrocopter pilot can steer his system to wherever he desires. This means technically that he can control the yaw angles x, y and z by himself.

However, with today's technology it is possible to measure and control all six degrees of freedom through the quadrocopter itself, without any external control of the pilot. It is therefore capable of staying stable in one place in the air, and the dream of an 'air nail' or 'air anchor' becomes true.

The following section describes briefly which sensors are needed to measure and control the desired degrees of freedom. The individual sensors are then discussed in their own subsections.

- Air pressure sensor: required for the degree of freedom in the z-direction.
- Compass: required for absolute angle measurement of the yaw angle (the gyro for the yaw angle measures only the angular velocity, as described in section 2.2).
- Global Positioning System (GPS): required for the two degrees of freedom in the x- and y-directions.

These four degrees of freedom can be added to the two degrees of freedom already measured and controlled with the basic components (the nick and roll angle), resulting in these six.

Before the explanation of the extension components, it should be pointed out that not every quadcopter pilot wants to equip his system with all sensors. Depending on the application, it may well make sense to use only the basic sensors, gyros and acceleration sensors. More technology support with additional controlled degrees of freedom also means that he has fewer degrees of freedom that must or can be controlled by hand.

2.4   Air pressure sensor

Measures the altitude and ensures its control (z-degree of freedom).

Figure 9: Air pressure sensor.

Air pressure decreases with increasing altitude. A reference pressure pushes on one side of a membrane and the air pressure pushes on the other side. If this changes, it distorts the membrane. This is measured by a strain gauge or a capacitive sensor. Figure 9 shows an air pressure sensor. The air flows through the hole in the middle in the direction of the membrane.

Not only a change in altitude results in air pressure changes in the sensor. Other factors, such as the air stream of the propeller or wind bursts, can also cause this. This must be taken into account during the assembly; placing the sensor directly on the control board with the hole facing upwards or downwards is appropriate. In order to shield the sensor from further disturbances, we can glue some foam or attach tape with a needle hole onto it. But when doing this it is important to ensure that the membrane is not damaged.

Control of altitude

There are several ways how the control of the height can be realized. Two possibilities will be presented here.

In the first variant, if the throttle stick is in the middle position, a zone is defined in which the quadrotor holds the reached altitude. It then adjusts the thrust by itself. If the throttle stick is above this zone, climb flight will be achieved, and if it is below, descent flight. This variant is the one that is most frequently used in the today's systems.

In the second variant the throttle stick must be moved so that the system passes into a climb flight. If a stick on the remote control is now moved, it holds the current altitude and controls the thrust by itself. If the gas is withdrawn, it turns into a descent flight. If afterwards the position of the throttle stick is increased again, the system returns to the climb flight. As soon as the height at which the lever was moved is reached again, it again holds that height. The control of the height therefore works metaphorically as an 'umbrella' above which the quadcopter can't fly out.

In both variants, the height control can be switched on and off with an additional remote control channel. This has the advantage that the model pilot can decide for himself when the quadcopter regulates the height by itself and when he has full control of the throttle stick and the thrust of the propellers

However, especially with RTF systems, it is very common that the support of the altitude control is permanently switched on and the model pilot always works with the function of the first variant. There are other variants for the implementation of the height control, but it would go beyond the scope of this booklet to explain them.

GPS is discussed later in this chapter, it provides is also information about the altitude. In contrast to the air pressure sensor, it indicates the absolute height, but with the inaccuracy inherent in GPS. Since the air pressure cannot be measured absolutely due to the constantly changing weather conditions, but only relatively, but has a higher resolution than GPS, both sensor principles are also combined in some systems. Good absolute accuracy and high resolution are then achieved. With many flight controllers, however, the value of the air pressure sensor is simply saved at start-up and the value is compared with the GPS. Since the weather conditions usually do not change noticeably during a short flight of a few minutes up to a maximum of half an hour, in most cases this is sufficient to reliably control the altitude even without GPS.

2.5 Sensors to measure the distance to the ground

As described, both the air pressure sensor and GPS have advantages and disadvantages with regard to altitude measurement. GPS only works to an accuracy of about 1 meter and an air pressure sensor does not measure the exact altitude either, since the air pressure is naturally subject to fluctuations. In addition, such a sensor must be calibrated each time it is started, as the air pressure is not constant at a certain altitude above sea level. The respective weather situation thus determines its absolute value. A high pressure weather situation ensures that all air flows and thus also possible clouds are pushed away, while a low pressure weather situation sucks them in accordingly. The combination of air pressure sensor and GPS does not always work in practice, because not every quadcopter is equipped with both types of sensors and GPS does not work for indoor flights either.

The exact height above sea level is actually not always important in practice. But the distance to the ground is particularly important, especially when the quadcopter is only hovering close to the ground or is about to land. In this case, other sensors are also used in modern flight systems, which are specialized in measuring the distance to the ground in a small area of a few centimeters to meters. Either triangulation sensors based on infrared are used here, or in some cases also ultrasonic sensors. These sensors can determine the distance to the ground to within a few centimeters. Depending on the application, they can cover an area of up to several meters. If they are used as an altimeter for take-offs and landings, this range is between one and five meters in common multicopter systems. The smaller the area, the more precisely the position can be recorded. For example, if a system can determine a height of up to two meters from the ground, a measurement accuracy of one centimeter is quite realistic.

Since triangulation sensors work with infrared light, i.e. optically, it is important that they are not soiled. Because they are always directed downwards, especially when landing, impurities can occur relatively easily. These should be wiped off with a handkerchief. Since triangulation sensors emit the infrared light with a narrow beam, even small bumps in the ground such as pebbles are measured. In order for the system to float stably at a defined height, several such sensors can be mounted on the underside and provide the average of several measuring points. In addition, the signal can also be averaged over time using suitable software filters. This problem is less pronounced when using ultrasonic sensors, because the ultrasonic signal is not transmitted in a narrow beam but with a larger opening angle. Thus, the distance is already averaged out by the measuring principle. Due to the non-optical measuring process, ultrasonic sensors are also less prone to contamination. However, both sensor systems are used in multicopters and are similarly reliable in practical application.

Automated take-offs and landings

If the drone used has ground distance sensors or an air pressure sensor and ideally GPS, fully automated take-off and landing maneuvers are possible. These are then also part of the simplicity in the operation of such aircraft and they also contribute significantly to the fact that today everyone is able to control them, because as with all flights with aircraft, the take-off and landing are also here the most critical flight phases. Of course, it must be assumed that the software supports these functions. The automatic start is then triggered, for example, with a corresponding button or switch, or by touching the corresponding button on the smartphone or tablet. Figure 10 shows the schematic diagram for the automatic take-off with a quadcopter, using ground distance sensors.

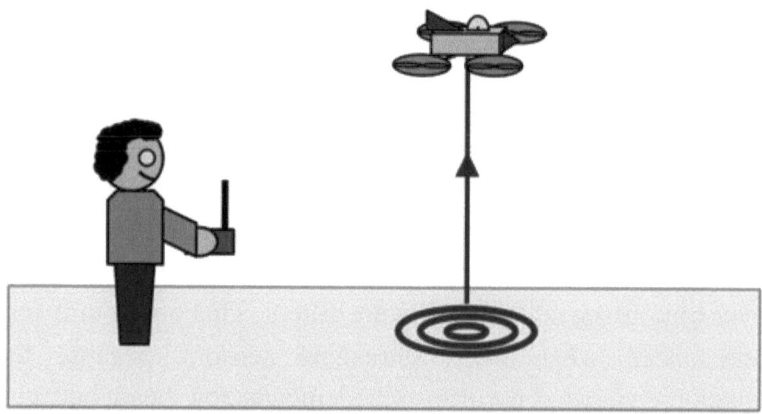

Figure 10: Schematic diagram for an automatic start with ground distance sensors.

The system then starts the motors independently and increases their speed until it lifts off the ground. Now the system constantly queries the ground sensor or the air pressure sensor and remains in a moderate climb until the target altitude for the take-off procedure is reached. This can be, for example, a height of 2 meters above the ground. A moderate climb means that at all times, in addition to the altitude itself, also the rate of climb is controlled. During the start phase this can be, for example, 0.5 m per second or less. Such a start procedure is basically possible without GPS, but it cannot be guaranteed that the system will drift to the side during the climbing phase, which would inevitably be the case, for example, with a little wind. However, if the system has a GPS, it can also correct the lateral drift during the climbing phase by briefly turning the motors on one side a little faster and the other a little slower. Some systems also use downward-facing cameras with image processing software during the start-up phase. These then try to climb as straight as possible with the help of the ground contours. Figure 11 shows a quadcopter, which is equipped with these sensors on the underside, and figure 12 shows a basic sketch of the measurement with a camera. This means that soil structures can also be recognized, so that with the appropriate software and contour-rich soil it is even possible to start without GPS on site.

Figure 11: Quadcopter with ground distance sensors on the bottom.

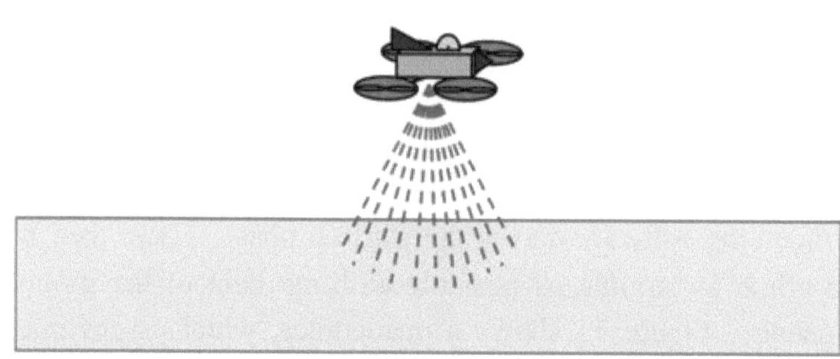

Figure 12: Basic sketch of a measurement of ground contours with a camera.

As soon as the target height is reached, the system will then hold it autonomously. Starting from this hovering position, the flight pilot can now carry out his own control commands.

The automatic landing can also be carried out fully automatically with this sensor support. As soon as the multicopter pilot presses the corresponding button or touches the button, the system must first measure the altitude. If only the command for a landing without a corresponding home function has been given, the system will initiate the descent on the spot. With an existing GPS or, as described above, with a camera pointing downwards, it is again possible to prevent sideways drift. The descent is again carried out completely autonomously at a defined rate of descent. It is at most possible that this descent rate is higher at greater distances from the ground and it is gradually reduced closer to the ground. The very small ground distances have to be treated in a very special way, because this is where the so called ground effect works. It can cause disturbing turbulence. In most systems, however, the software is so refined that the quadcopter touches the ground gently at the end and switches off the motors automatically.

2.6 Compass

Measures the absolute angle in the yaw axis and ensures its control (yaw degree of freedom).

The gyro in the yaw axis only measures the angular velocity (and not the angle itself). When the yaw stick of the remote control is now held in the middle position, that means angular velocity 'zero' for the control electronics board or the flight controller. This also means that the yaw angle will not be changed. The 'nose' (i.e. the front, as far as one can speak of 'nose' and 'front' with regards to quadrocopters) always remains in the same place. This is true in theory at least.

In practice, the gyro measures an angular velocity of zero even if this is in fact not exactly zero. Because of temperature effects this can't be prevented. A gyro calibration before the flight does not help either.

Thus, the 'nose' will turn away more or less slowly, depending on the quality of the calibration or trimming. The model pilot compensates for that in the same way as with helicopters, namely by a counter-movement of the yaw stick or the trimming.

If it is desired that the quadrocopter always holds the nose at the same place at the yaw stick position zero, an additional sensor is needed, namely a compass. This will no longer measure a relative motion to the ground, but the absolute orientation.

Many electronic compasses operate on the basis of Hall sensors. The Hall effect is the occurrence of an electrical voltage in a current-carrying conductor which is placed in a magnetic field. The earth produces a magnetic field between the north and south poles, which is also responsible for ensuring that the compass needle always aligns to the north. The word 'always' should to be treated with caution. If you are in a closed room with metal in the walls or in an area with power cables or iron-rich rocks then the compass needle can be deflected.

The electronic compass is also affected by such disturbances. It is therefore important that they are well shielded or suppressed by other appropriate measures. So the compass should be mounted as far away as possible from the control electronics, the brushless control or piezo buzzer. Because of metal in the walls, problems with incorrect measurements are often also reported with indoor flights. Since the built-in compass moves on the nick and roll movement together with the quadrocopter, it must be designed in a tilt-compensated manner, otherwise it will also result in errors.

Control of the yaw angle

As with the realization of the height control, there are also different options for the regulation of the absolute yaw angle. A simple solution is similar to one described there. If the yaw stick is in the center position, or it deviates only slightly from it, then the angle control is active together with the compass and the current absolute angle is kept. If the stick position is sufficiently far enough left or right, it is regulated using the gyro sensor as in the configuration with the basic components. This means that the farther left or right of the center position of the stick is, the faster it turns the nose.

The compass and the acceleration sensor, which measure the yaw angle, respectively the nick and roll angles, have an interesting commonality. With both types of sensors, the angles are measured in the world coordinate system. In other words, using those sensors, the angular position of the quadrocopter is always measured relative to the Earth. Thus, the sensors must also be based on earth- related measurement principles. In the case of the compass it was mentioned above that it measures the magnetic field between the north and south poles of the earth. With the acceleration sensor it can be seen in figure 7 that the gravitational acceleration acts as a reference there. It always points to the center of the earth, so it is also earth-related.

Without the GPS described in the next chapter, all three angle degrees of freedom can be controlled with the sensor types discussed above: nick, roll and yaw. Furthermore, the z-axis (gas) can also be controlled using the air pressure sensor. In total that corresponds to a control of four of the six degrees of freedom.

Figure 13: Compass and electronics board.

Care free control, headless control or orientation-independent control

As soon as a quadcopter is equipped with a compass, the software or the flight controller can allow a completely different type of control. You can then switch to a new mode on some systems. With this, the functions 'Nick' and 'Roll' always point in the same direction from the pilot's point of view, regardless of the current orientation of the quadcopter. As soon as the yaw stick is operated, the built-in compass continuously calculates how its own orientation changes. From this it converts the 'Nick' and 'Roll' control commands and controls the system in such a way that it always appears the same from the pilot's point of view. In some systems, this function is called, for example, 'care free' or 'headless' or similar, if it is available. If you have mastered this mode and the software allows such maneuvres, you can, for example, make the quadcopter rotate continuously around the vertical axis and still rotate in the desired direction with nick and roll. It is important that the pilot does not turn away together with the quadcopter, but always remains in the same place and looks in the same direction. Otherwise he runs the risk of becoming disoriented himself. In this case, the advantages of this mode can quickly become a disadvantage. If you want to master your quadcopter in all flight positions and orientations, you should consider for yourself whether you want to use this mode for the exercise. When practicing, it is especially important at the beginning to hold the quadcopter in the correct orientation with the yaw stick. But that is precisely what is no longer necessary with the 'care free' or 'headless' function. A quadcopter pilot should be able to use the yaw mode. But in any case, this function is an interesting addition to the quadcopter flight.

3.   GPS

3.1   GPS sensor principle

With the sensors discussed so far, all three of the angular degrees of freedom and the Z-axis (gas, height can be regulated. This corresponds to a control of four of the six degrees of freedom according to figure 8.
1. Nick: with gyro and acceleration sensor (generally inertial measurement unit IMU)
2. Roll: with gyro and acceleration sensor (generally inertial measurement unit IMU)
3. Yaw: angular velocity only: gyro; absolute angle: gyro and compass
4. height (Z): air pressure sensor, air pressure sensor with GPS or ground distance sensors.

GPS, Global Positioning System, has affected the everyday life of people for many years. Electronic navigation systems are now standard in many vehicles. There are simple technical solutions, providing a GPS signal on a single chip.
Soon after the first quadcopters were able to fly reliably, the advantages in the combination between them and the GPS were recognized. So the drone technology was soon expanded with control electronics, which also included GPS receivers. The combination of both technologies today enables various areas of application in model making and technology.

In practice, one is most likely to want to bring a camera into the air. The combination with GPS is particularly interesting for the creation of photos and films, as it can be used to stabilize the quadcopter at almost any location. In this way, images can be created from almost every possible perspective. In another case, the pilot may be less interested in the photo flight, but simply enjoy the technical possibilities and the fact that his quadcopter can stop like an 'air dowel' or fly to preprogrammed waypoints. However, there are many other conceivable and feasible applications now and in the future.

There are several good reasons why the two technologies GPS and quadcopter are used together.

On the one hand, it is the time of their maturity that connects them. The first unmanned remote controlled quadcopters were built in the 1990s. However, it was only the progressive miniaturization of electronics, microcontrollers and the development of brushless DC motors that gave them their breakthrough in the early 2000s. GPS has been known since the 1970s. However, it only became interesting for civil use in 2000, when the artificial deterioration of the satellite signals was switched off. From this point on, every receiver was able to determine its position on earth with an accuracy of better than 10 m. Before that, this was only possible to achieve an accuracy of around 100 m.

On the other hand, the quadcopter is an aircraft that is able to fly to any point in the air with the simplest mechanics and some modern sensor and microprocessor technology. With its help, it is no longer possible to just head for one point on the floor, but to do so in the third dimension. While GPS represents the three-dimensional sensor, the quadcopter is to a certain extent the three-dimensional actuator that suits it.

Basics

The GPS consists of several satellites orbiting the earth. Their exact position is known at all times. Each satellite is equipped with an atomic clock and sends out a signal at certain times. That signal also contains its identification. The ground station, a navigation device or a chip on the quadrocopter, measures the running time of the transmitted signals. This is the time which passes between the emission of the signals through the satellites and their reception. In this way it can determine its own position.

The running time multiplied by the speed of light is the distance from the satellite. To locate a particular place on earth, the signals from at least three satellites should be visible at the same time. To measure the exact duration, the ground station would also have to be equipped with an atomic clock, so that all the devices use the same time base. Because that would be much too expensive, in practice such a clock is not installed at the ground station. In order to nonetheless determine the position, the position signal of a fourth satellite as shown in figure 14 is needed. Thanks to special algorithms, the ground station can then calculate the longitude and latitude by itself.

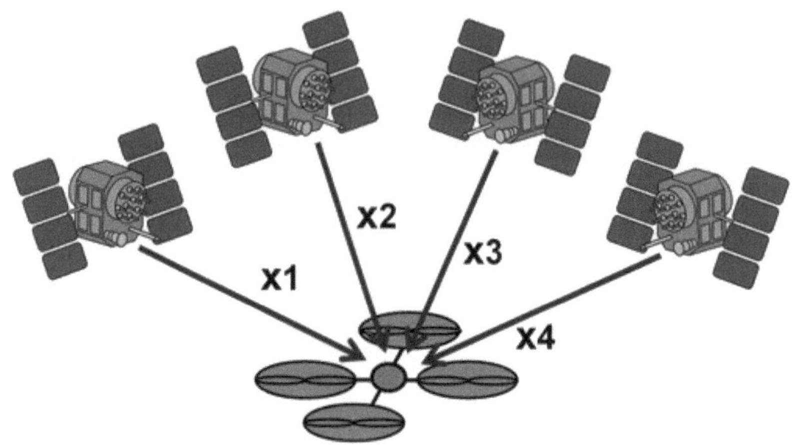

Figure 14: Four GPS satellites and a receiver on a drone.

Each satellite transmits its current position and the timebase. So the four satellites allow a position determination in the three-dimensional space. In addition to the position on earth, also the altitude is measured. The fourth satellite makes this possible for the above-mentioned reason, and it also provides the exact time. The time is also considered the fourth dimension. For this reason, every GPS receiver can access the exact time. Thus, as soon as enough satellites are visible, it is actually also something like a virtual atomic clock.

Number of satellites

For the correct function of the transit time measurement, the satellite signals must not be reflected on the earth's surface before they reach the receiver. In this case, the distance covered by the waves would be greater than the actual distance. For this reason there are sometimes incorrect measurements between rows of houses or in the vicinity of high mountains.

To ensure that at least four satellites are always visible in terms of direct signal reception, the fully developed GPS system consists of at least 24 satellites. They circle the earth on six orbits, each rotated by 60° to each other. In fact, over 30 satellites are active on these orbits. A failure of a single one cannot jeopardize the function. In this way, there are usually not just four, but up to ten or more satellites at the same time above the switch-off angle, i.e. the angle above the horizon, above which the signals can be received without reflection.

The more satellites are visible, the more precisely the receiver can determine the position. If more than four satellites are visible, the receiver does an optimization calculation. Its algorithm finds the optimum position with all reliable GPS satellites. In practice, a position determination is carried out with at least eight to ten satellites that are visible at the same time.

Figure 15 shows an expanded representation of the visible satellites and the signal strength received in each case. Since they all send out a clearly identifiable signal, which also includes the position, the navigation device can create a map of where they are. The intensity of the signals received from each of the satellites can also be measured. Those satellites that are directly above the receiver, which are in the middle of all directions, are the least far away. This is why they are received with the greatest signal strength. The further away you are from the center, the weaker the signals are. Based on the quality and signal intensity of these signals, and the angle above the horizon, the receiver software then decides whether or not the data sent by the visible satellites will be included into the position calculation.

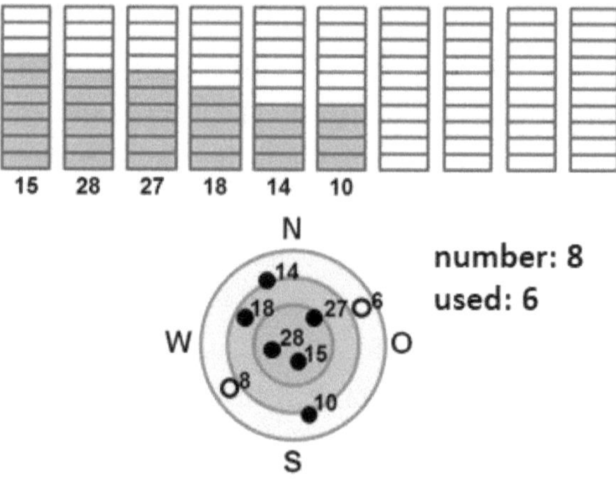

Figure 15: Illustration of the signal intensity and position of the satellites.

GPS for indoor

In principle, electromagnetic waves can also be transmitted into a building. After all, this is how every radio with an extendable antenna works. Due to the low transmission power of the GPS signals, however, a direct line of sight is usually necessary.

But the signals are reflected inside the building and thus falsify the position measurement. An exact indoor measurement is therefore not possible due to the reflections. Thus, other sensor principles are used. This can be, for example, three ultrasonic sensors placed in the room. The electronics of the drone then evaluate this to determine the position, similar to a GPS signal.

Other technologies

In addition to GPS, there are also other positioning systems that work on the basis of satellites. Since the positioning systems are also used militarily, it is of strategic importance for larger countries or alliances of states to operate their own systems. GLONASS is the Russian system and Europe also has a positioning system, namely Galileo. China, India and Japan also have such systems. The modern, highly integrated GPS chips take advantage of the availability of these positioning systems. Some of them also receive this data and evaluate it using complex software algorithms. In this way, they calculate a sensor and data fusion, and thus improve the accuracy of the position determination with the inclusion of all positioning systems. So if GPS is mentioned in this book, then this is actually only to be understood as a generic term, because in practice the chips include not only GPS but also the other positioning systems in the calculations.

3.2   GPS in drones and quadcopters

Gives the world's coordinates in longitude and latitude and allows the quadcopter to be regulated in the x- and y- axes (x and y degrees of freedom).

Chapter 2 has been carried out so far that the quadcopter with gyro and acceleration sensors as well as compass and air pressure sensor could be controlled in all three solid angles and the height (Z). That is a control in four of the six degrees of freedom.

If you want the quadcopter to be able to autonomously fly to any point in the airspace, you need the GPS for the two other degrees of freedom. In practice, these two coordinates correspond to the longitude and latitude. In order to remain compatible with the coordinate system that applies to a body in space, the terms X and Y are still used here. In this way, all six degrees of freedom of the system are controlled independently. It then becomes a 'nail in the air', so it can, for example, remain in one place without external control. It was described above that GPS also provides the height (Z) in addition to the coordinates for X and Y. At first glance, it seems as if there is a duplication between GPS and the air pressure sensor. In reality, the GPS accuracy of a few meters is not good enough for reliable altitude control. On the other hand, the air pressure sensor has the property that it has to be recalibrated again and again depending on the weather conditions with high or low pressure areas. Thus, similar to determining the angle using a combination of gyro and acceleration sensor, it is also the combination of the good properties of GPS and air pressure sensor that achieves a good result when measuring altitude. Often the rough altitude movements are calculated with GPS, and the fine adjustment and control are carried out with the air pressure sensor.

At this point it should be pointed out that a system operated in this way, which is regulated in all degrees of freedom, is no longer a pure model aircraft in the usual sense. There it is part of the thing that you can steer it in different directions with the remote control. However, a system regulated in all six degrees of freedom relieves the user of all control tasks, so he no longer has to move the sticks on his remote control transmitter. Due to the increasing spread of miniaturized GPS receivers to the size of a chip, there are more and more such unmanned and autonomous systems in the airspace. Many state governments issue guidelines for their use. This is also necessary, because these quadcopters also represent a potential safety risk. As a minimum requirement, the user must be able to guarantee that he has visual contact to his quadcopter at all times and that he can take full control again at any time via remote control.

The legislation is constantly being adapted here. The technology of GPS-controlled quadcopters is still young in the first quarter of the 21st century. What was unimaginable just a few years ago is already a technical standard today. Many state governments will therefore have to continuously enact further laws, be it that the pilot must at least be registered or even that he has to complete a training course that enables him to fly with the drone. Due to the large number of commercially available drones and the sometimes controversial discussions of politics and society, it is quite possible that this type of control for private individuals will at some point be completely banned or only permitted under strict conditions. From today's perspective, however, it is certain that the future legislation will be tightened rather than relaxed when it comes to the construction and operation of drones. In the further course of this chapter, only the technology of the GPS-controlled quadcopter will be discussed. However, it is up to every quadcopter pilot to familiarize himself with current and constantly changing laws and to comply with them

Technics

Small devices with GPS are booming in the first quarter of the 21st century and almost everyone today is probably the owner of a GPS receiver in some form, for example that of a smartphone. This only became possible through the advancing miniaturization of chip technology. The receivers can now be accommodated on a single chip. So the step towards the solution for a quadcopter was easy to take. Figure 16 shows such a solution. It is the receiving module of a quadcopter. Basically nothing more than the GPS receiver module is visible on the circuit board. It is a highly integrated electronic component.

Figure 16: GPS receiver on a chip.

Figure 17 shows in the middle a GPS antenna belonging to the receiver, which is already mounted on a quadcopter. It is designed as a so-called patch antenna. This type of antenna has a strong directional effect on the top of the plate. So it can amplify signals from this direction very well. Typically it consists of a plate with half the wavelength in length. Underneath there is a dielectric, i.e. a non-conductive layer. Then another conductive layer follows. This acts as a reflector upwards and shields downwards. In this type of installation, all electromagnetic waves that come from above are optimally amplified. In flight, this is exactly the direction from which the signals sent by the satellites come.

Figure 17: Patch antenna of a GPS receiver mounted on a quadcopter.

Figure 18 shows a fully integrated GPS module including a compass. It can be connected directly to common flight controllers. So if the controls of all sensors are switched on, i.e. gyros, acceleration sensors, air pressure sensors, compass and GPS, the quadcopter will stand still more or less rooted in the air. All six degrees of freedom of a body in space are then controlled autonomously.

Figure 18: Combined GPS and compass module.

4. Other sensors

4.1 Sensors for the detection of obstacles

Quadcopter manufacturers are bringing more and more sophisticated solutions for obstacle detection to the market. On the one hand, they are the increasingly better and more reliable sensor systems and, on the other hand, the increasingly complex software systems make this possible. Either the triangulation sensors based on infrared or ultrasonic sensors that have already been discussed in connection with the ground distance sensors are used. In some systems, the software also evaluates the images from the cameras in order to identify possible obstacles. There are also major differences in the arrangement of the sensors. With some quadcopters only a sensor to the front is attached, with others, if they are available, the floor sensors are also evaluated, and again with others there are also sensors that are oriented to the rear and to the side. The distance sensors are often supplemented with small cameras on several sides. Of course, the sensors used also have an influence on the quality of the obstacle detection. Systems with forward-facing sensors can only detect obstacles in forward flight and which are in front of them in this flight direction. If sensors are mounted on all sides, obstacles can also be evaluated on all sides and in every direction of flight.

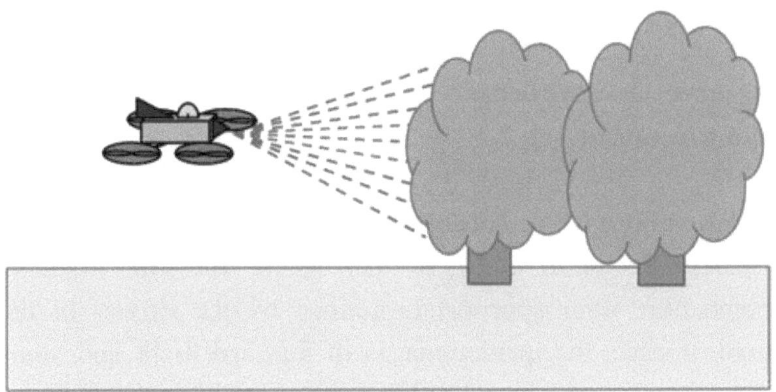

Figure 19: Basic sketch of obstacle detection, here with sensors facing forward.

The software for this is of course a very important component here too. It is not that easy for an automated flight system to recognize obstacles without a doubt. In most cases, these do not have a well-defined shape, but consist of corners, curves or deepenings. It becomes even more difficult with trees and bushes, because depending on the location, the infrared light from a triangulation sensor, for example, can either hit an outer leaf or it only measures free space deep into the plant, i.e. no distance to an object at all. This makes correct distance detection much more difficult. Sophisticated algorithms are used here, which try to calculate the true distance to the obstacle as well as possible. Often the camera itself is used for obstacle detection in addition to its actual task of photo flight. In this case, software elements from modern image processing are used. This makes it possible to recognize entire structures of the obstacles and, together with the infrared or ultrasonic sensors, to determine their distance. Often there are sensors in combination with small cameras on several sides of the quadcopter, for example also downwards. Thus, today's software systems including image processing algorithms can recognize the environment quite well from the data obtained from the camera.

Actions taken when detecting obstacles

Obstacle detection alone is only useful for safe flight in conjunction with appropriate actions by the drone. In the simplest case, the quadcopter is in forward flight and stops when there is an obstacle. Then the pilot has to decide for himself what the drone shall do. It can turn or fly backwards or swerve to the left, right or even upwards. In systems with more complex algorithms, the system evades independently and flies around the obstacle, so to speak, and then switches back to forward flight. Different scenarios are possible, for example it can move upwards or to the side. The GPS and the electronic compass are also included here, since the direction of flight has to be correct in the end. If the appropriate sensors are available, the above also works sideways and backwards.

However, such automatic evasive maneuvers already require a complex software structure and the exact execution of the corresponding function must also be preconfigured in individual cases. The obstacle detection actually works in a distance range between about half a meter and a few meters and corresponds to the range of the ultrasonic and infrared triangulation sensors used. It can also be switched off on most systems, if present. Detection within a few meters also means that the system can still react and brake in good time. When obstacle detection is switched on, the speed is therefore typically reduced somewhat compared to normal operation, for example by half. Typical values are around 10km/h to over 30km/h flight speed, compared to around 20km/h to over 60km/h with obstacle detection being switched off. The values vary depending on the quadcopter system, in particular higher speeds are achieved with larger sensor ranges. Of course, this also goes hand in hand with a combination with larger and more powerful motors and often with more powerful microprocessors. This means that more complex and reliable software algorithms can also be used for obstacle detection.

4.2 Sensors for tracking or positioning without GPS

Tracking of moving objects

The built-in camera in combination with ultrasonic or triangulation sensors based on infrared can, with the right image processing software, do a lot more than just detect obstacles. Since these systems are basically also able to recognize moving objects such as people, they can also be used to track them. If the quadcopter is equipped with the appropriate sensors and software algorithms, the relevant image section can be selected with the moving object on the smartphone or tablet. Then you can also select the distance between the object and the quadcopter within the permissible measuring range of the sensors. In the actual tracking mode, the quadcopter then follows the moving object and always keeps the set distance constant.

The software also ensures that the camera with its gimbal is always aimed at the object. It is even possible that the quadcopter does not follow the object from behind, but from the side, from above or even from any angle that can be selected. This means that a large number of automated applications are possible. Examples of this are sports events with tracking and making film recordings of runners, skiers, mountain bikers, or even parachute pilots. Or a model car or a model boat is followed on its circuit. And of course it is also possible to track a model airplane or a quadcopter with a quadcopter, as long as it is fast enough. Object tracking can, however, also take place in other ways than with image processing. If the object to be tracked has a GPS system itself and transmits its current position to the quadcopter via radio, the tracking can also be carried out at any angle and distance via the GPS system itself, without the use of image processing. And here, too, the camera can be aimed precisely at the object with the gimbal.

Outdoor and indoor positioning with floor sensor and camera

If the system has sensors pointing downwards such as infrared or ultrasound and also a camera pointing downwards, the corresponding software can do more than just detect obstacles.

The camera can then detect unevenness or contours of stones or floor slabs and possibly even blades of grass on the meadows and thus has a relative localization. This option is particularly helpful for indoor flights. GPS does not work there due to restrictions from physics. Thus, this function can be used with a correspondingly contoured ground structure and it is then possible, even without a GPS signal, for the quadcopter to hover independently and in all coordinates on the spot, without the flight pilot is making corrections with the remote control or the smartphone.

5   Literature

[1] Integrierte Navigationssysteme. Sensordatenfusion, GPS und Inertiale Navigation, Jan Wendel, Verlag Oldenbourg 2007, ISBN: 978-3-486-58160-7

[2] Global Positioning Systems, Inertial Navigation, and Integration, Mohinder S. Grewal, Lawrence R. Weill, Angus P. Andrews, Verlag: John Wiley & Sons, Inc. 2007, ISBN: 978-0-470-04190-1

[3] Buchi, Roland, et al. "A remote controlled mobile mini robot." *MHS'96 Proceedings of the Seventh International Symposium on Micro Machine and Human Science.* IEEE, 1996.

[4] Moderne Fernsteuerungen für RC-Flugmodelle, Manfred-Dieter Kotting, Verlag für Technik und Handwerk GmbH, ISBN: 978-3-88180-780-7

[5] Büchi, Roland."Brushless-Motoren und–Regler. 1." Auflage, Baden-Baden: Verlag für Technik und Handwerk neue Medien GmbH (2013).

[6] Selbstbau von Brushless-Aussenläufer-Motoren, Heinrich Hilgers, Neckar-Verlag 2006, ISBN 978-3-78830-683-0

[7] Regelungstechnik, Einführung in die Methoden und ihre Anwendung, Otto Föllinger, Hüthig Verlag 2008, ISBN: 978-3-7785-2970-6

[8] Büchi, Roland. *Radio control with 2.4 GHz*. BoD–Books on Demand, 2014., ISBN 978-3732293407